FLORA OF THE GUIANAS

Edited by

A.R.A. Görts-van Rijn
&
M.J. Jansen-Jacobs

Series A: Phanerogams
Fascicle 20

10. ARISTOLOCHIACEAE
(C. Feuillet & O. Poncy)

including
Wood and Timber
(B.J.H. ter Welle & P. Détienne)

1998

Royal Botanic Gardens, Kew

Contents

© The Trustees of The Royal Botanic Gardens, Kew.
ISBN 1 900347 44 X

10. ARISTOLOCHIACEAE

by

CHRISTIAN FEUILLET[1] & ODILE PONCY[2]

Woody or herbaceous vines or less often perennial herbs or shrubs, sometimes arising from tubers or from fleshy roots; nodes 3-lacunar. Leaves alternate, simple and mostly entire, sometimes lobed; stipules wanting, a pseudostipule present in the leaf axil in some South American species. Flowers solitary, or in terminal or lateral racemes or cymes, perfect, actinomorphic or zygomorphic, mostly epigynous, otherwise hemi-epigynous or perigynous, often smelling like rotting meat; calyx synsepalous, tubular at least below, actinomorphic and 3-lobed to zygomorphic and 3-lobed to 1-lobed or not lobed, often large and colourful; petals wanting, much reduced or well developed (*Saruma*); stamens 4-many, anthers dithecal and tetrasporangiate, extrorse; gynoecium of 4-6 carpels; ovary 4- to 6-locular, or with 4-6 incomplete partitions, placentation axile or parietal; ovules numerous, crassinucellar and mostly anatropous; endosperm development cellular. Fruits usually capsular, rarely follicular or indehiscent; seeds with abundant, oily endosperm; embryo very small, sometimes undifferentiated, basal.

Distribution: Throughout the world in tropical and temperate regions; an estimated 650 species in 6-7 genera (3 monospecific), most of the species (about 550) belong to *Aristolochia*, and most of the remaining (about 75) to the north temperate genus *Asarum*; in the Neotropics probably about 250 species in 3 genera; in the Guianas 18 species in one genus.

Pollination: The flowers are usually pollinated by flies attracted by the strong smell of the flowers.

LITERATURE

Duchartre, P. 1854. Methodicae Divisionis Generis Aristolochia. Ann. Sci. Nat., sér. 4, 2: 29-76.

[1] Department of Botany, Smithsonian Institution, Washington, D.C., U.S.A.

[2] Laboratoire de Phanérogamie, Museum National d'Histoire Naturelle, Paris, France.
  Drawings made by G. Chypre.

Duchartre, P. 1864. Aristolochiaceae. In A.P. De Candolle, Prodr. 15(1): 421-498.

González, F. 1990. Aristolochiaceae. In Flora de Colombia 12: 1-184.

González, F. 1991. Notes on the systematics of Aristolochia subsect. Hexandrae. Ann. Missouri Bot. Gard. 78: 497-503.

González, F. 1994. Aristolochiaceae. In G. Harling & L. Andersson, Flora of Ecuador 51: 3-42.

Grenand, P., C. Moretti & H. Jacquemin. 1987. Pharmacopées traditionnelles en Guyane. ORSTOM Mémoires 108. Aristolochiaceae, p. 144-149. Paris.

Hoehne, F.C. 1927. Monogr. Aristol. Bras., Mem. Inst. Oswaldo Cruz 20(1): 1-111.

Hoehne, F.C. 1942. Aristolochiaceas. In F.C. Hoehne, Flora Brasílica 15(2): 1-141, 123 tab.

Malme, G.O.A. 1904. Beiträge zur Kenntnis der südamerikanischen Aristolochiaceen. Ark. Bot. 1: 521-552, t. 31-33.

Masters, M.T. 1875. Aristolochiaceae. In C.F.P. von Martius, Flora Brasiliensis 4(2): 77-114, tab. 17-26.

Pfeifer, H.W. 1966. Revision of the North and Central American hexandrous species of Aristolochia (Aristolochiaceae). Ann. Missouri Bot. Gard. 53: 115-196.

Schomburgk, M.R. 1849. Reisen in Britisch-Guiana in den Jahren 1840-1844. Vol. 3. Flora. 787-1260. Leipzig.

1. **ARISTOLOCHIA** Linnaeus, Sp. Pl. 960. 1753; Gen. Pl. 410. 1754. Lectotype: A. rotunda Linnaeus

Howardia Klotzsch, Monatsber. Königl. Preuss. Akad. Wiss. Berlin 1859: 584, 607. 1860, non Wedd. 1854.
Type: H. ringens Klotzsch

Aromatic woody or herbaceous vines, roots often tuberous; stem of several species with thick, fissured cork. Leaves: with or without axillar pseudostipules; petioles long, twisting; blades entire, 3-lobed in 2 species, base usually cordate, venation prominent beneath, main veins pedate at base, tertiary veins forming a tight net. Flowers solitary or in racemose or cymose inflorescences, axillary or cauliflorous; flowers resupinate, as a result of twisting of the inferior ovary; perianth composed of 3 fused sepals, zygomorphic, slightly curved to tightly S-shaped, its 3 parts easy to distinguish: the basal, inflated utricle, the narrow or trumpet-shaped median tube, the distal part, usually 1- to 3-lobed, very variable in shape, size and colour; ovary 6-celled, long and narrow, barely thicker than the pedicel, apex forming a hexandrous gynostemium with 6 anthers and 6 stigmatic lobes in the utricle. Capsules 6-locular, cylindric, with dorsal keel on each cell; dehiscence septicidal, starting from the peduncle and incomplete, dehisced fruit basket-like; seeds numerous, generally flat, piled in each cell, often winged, seed coat slightly verrucose, adaxially with a longitudinal prominent raphe.

Distribution: Tropical and subtropical regions of the world; only a few temperate species. In the Guianas in dense primary forest and secondary vegetation or disturbed areas.

Phenology: Flowering is erratic among the species growing in the forest in the Guianas. All species are rarely collected (which does not permit to draw trustable conclusions about the phenology of the taxa growing in open vegetation).

Subdivisions: All known South American species belong to sect. **Gymnolobus** Duchartre subsect. **Hexandrae** Duchartre as traditionally accepted since Duchartre (1864). González (1991) divided this subsect. in 2 series:
(1) ser. **Thyrsicae** González, characterized by the presence of an abscission zone at the base of the petiole, the structure of the inflorescence, the comb-like shape of the wall of the dehiscing fruit, and 2-winged seeds;
(2) ser. **Hexandrae** splitted into 2 subseries **Anthocaulicae** and **Hexandrae**.
According to this system, *A. consimilis* Masters and *A. paramaribensis* Duchartre should fall into ser. **Thyrsicae**. As far as their fruits are known, the other Guianan species belong to ser. **Hexandrae** subser. **Hexandrae** (non cauliflorous) and subser. **Anthocaulicae** González (cauliflorous).

Economic uses: Several species are in cultivation in tropical and temperate areas as greenhouse or garden plants prized for their flowers. Often *Aristolochia* species are used in folk medicine.

Note: The alleged property to ease child birth explains the scientific name of the genus, "Aristo-lochia" = best-delivery" in Greek, as well as one of its English vernacular names: "birthwort". Medicinal data and vernacular names were taken either from Grenand et al. (1987) or from cited specimens.

Vernacular names: French Guiana: bukuti (Wayampi, Palikour), liane amère (Creole; also applied to *Tynospora* of the Menispermaceae).

KEY TO THE SPECIES

1 Leaves 3-lobed · 2
  Leaves not lobed · 3

2 Pseudostipules lacking; lateral lobes of leaves perpendicular to middle lobe;
  superior lobe of perianth triangular and emarginate at apex · · · · · · · · · ·
  · *16. A. surinamensis*
  Pseudostipules present; lateral lobes of leaves falciform, pointing partly
  forward; superior lobe of perianth ovate, with a long apical filament at
  least 10 cm long · *17. A. trilobata*

3 Leaves peltate · *13. A. peltatodeltoidea*
  Leaves not peltate, or very slightly so · 4

4 Plant cauliflorous and leaves wider than long · 5
  Plant not cauliflorous or, if cauliflorous, then leaves longer than wide · · · 7

5 Limb of perianth not spreading, but at first reflexed backwards to the utricle
  and apically coming forward, but finally recurved outwards and forming
  a horse-shoe-shaped cavity around the mouth of the tube; fruit more than
  6 cm long, external wall thick, woody · · · · · · · · · · · · · · · *15. A. stahelii*
  Limb of perianth spreading; fruit similar or thin-walled · · · · · · · · · · · · 6

6 Upper part of the limb wider than long, apex rounded; fruit more than 6 cm
  long, external wall thick, woody · · · · · · · · · · · · · · · *4. A. daemoninoxia*
  Upper part of the limb longer than wide, apex acute; fruit up to 6 cm long,
  external wall thin, chartaceous · *1. A. bukuti*

7 Perianth with only 1 superior lobe (with small inferior lobe in *A. cremersii*)
  · 8
  Perianth with 2 lobes (2 lateral lobes, or 1 superior and 1 inferior lobe) · ·
  · 15

8 Perianth lobe hanging, like a scallopped tape ·
  · *18. A. weddellii* var. *rondoniana*
  Perianth lobe erect · 9

9 Inflorescences cauliflorous or axillary racemes · · · · · · · · · *3. A. cremersii*
  Flowers solitary or arranged in axillary inflorescences · · · · · · · · · · · · · 10

10 Flowers solitary; inside wall of dehiscing fruit not comb-like · · · · · · · · 11
  Flowers in axillary inflorescences; inside wall of dehiscing fruit comb-like
  · 14

11 Perianth large, utricle 3 x 1.5 cm, larger than the lobe · · · · · *11. A. mossii*
  Perianth with the utricle smaller than the lobe · · · · · · · · · · · · · · · · · · 12

12 Perianth lobe triangular, without fimbriae, but with dark violet spots · · · · ·
· 7. *A. guianensis*
Perianth lobe with fleshy, dark purple fimbriae · · · · · · · · · · · · · · · · · 13

13 Perianth lobe triangular, acute, marginally reflexed · · · · · 10. *A. leprieurii*
Perianth lobe distally ovoid, with acuminate apex · · · · · · · · 14. *A. rugosa*

14 Perianth lobe with terminal filiform appendage; seed wings apically
truncate, seed base truncate · · · · · · · · · · · · · · · · 12. *A. paramaribensis*
Perianth lobe triangular with acute apex; seed wings apically rounded, seed
base cordate · 2. *A. consimilis*

15 Perianth with 2 lateral lobes · 9. *A. iquitensis*
Perianth with superior and inferior lobe · 16

16 Utricle less than 1.5 cm long, superior lobe linear, twisted · · · · 6. *A. flava*
Utricle more than 2 cm long · 17

17 Superior lobe deeply 2-fid, up to 5 cm long · · · · · · · · · · · · 5. *A. didyma*
Superior lobe paddle-shaped, 1.5-3 cm long, wider in its distal part · · · · ·
· 8. *A. hians*

1. **Aristolochia bukuti** Poncy, Bull. Mus. Hist. Nat. (Paris), sér. 4, 10: 337, tab. 1. 1989 ("1988"). Type: French Guiana, Arataye R., Poncy 536 (holotype P, isotypes CAY, US).

Woody vine; stem 1.5-2 cm diam., bark very thick and fissured; branchlets rather thick, 3-4 mm diam., indument caducous; pseudostipules lacking. Leaves: petiole thick, twining, 5-7 cm long; blade chartaceous, widely triangular, 11-16 x 15-20 cm, glabrous and shining above, greyish and pubescent beneath, apex acute to slightly acuminate, base slightly cordate; lateral veins marginal at base. Inflorescence cauliflorous, a dense cluster of 1-2 cm long racemes, each bearing ± 5 flowers with small scaly bracts; pedicels thin, 3-3.5 cm long, excluding the ovary. Flowers: ovary 2-2.5 cm long; perianth whitish, glabrous, outside with protruding violet brown veins; utricle ovoid, 2-3 x 1.5 cm, with a cup-shaped, annular, 1 mm wide appendage at base; tube 2 x 0.5 cm; limb suborbicular, surrounding the throat except for a ventral sinus, 5-6 x 4-5 cm, apex acute, inside brown with golden speckles near the margin, the colour fading to yellow at middle, throat yellow. Fruit ca. 6 cm long, external wall thin, chartaceous.

Distribution: Only known from French Guiana; in forests on slopes (FG: 5).

Specimens examined: French Guiana: near Saül, Mori 21637 (CAY, NY, P, US); Camopi, Oldeman 2622 (CAY, P); Cabassou, Prévost 604 (CAY, P); Mt. Bellevue de l'Inini, de Granville 7968 (CAY, P).

2. **Aristolochia consimilis** Masters, Bull. Misc. Inform. 1906: 7. 1906. Type: Guyana, Demerara R., Jenman 6916 (holotype K, isotypes BRG, K, NY, P-JUSS, B probably not extant, photo COL, F, G, MO, NY, US).

Woody vine; stem 15 x 10 mm in cross section, bark thick and fissured; branchlets with fine, short and sparse trichomes; pseudostipules lacking. Leaves: petiole twining, about 2 cm long; blade chartaceous, elliptic, 7-11 x 4-7 cm, pubescent when young, becoming glabrescent except on the veins and dull above, beneath pubescent, apex acute to acuminate, base deeply cordate. Inflorescence axillary, hanging, 1-2(-3) together; racemes 2-4 cm long, occasionally up to 10 cm, about 1 cm between flowers; bracts small, leafy, linear to triangular, 6-10 x 2-4 mm; pedicels ± 5 mm long not longer than the ovary, pubescent. Flowers: ovary ± 7 mm long, pubescent; perianth pilose with multicellular hairs, dark purple with green veins outside; utricle ovoid, 7 mm long; tube 10-15 mm long, broadening into a flattened mouth; lobe triangular, about 15 mm long and wide, apex acute, inside yellow. Fruit (4-)5-6 cm long, glabrous, with a longitudinal ridge along the middle of each carpel, inner walls membranaceous, comb-like; seeds cordiform, 5 x 5 mm, surrounded by 2 parallel, papyraceous wings, with rounded apex, 7-8 x 12-15 mm including wings, base cordate.

Distribution: Guyana (NW region) and French Guiana (GU: 7; FG: 1).

Selected specimens: Guyana: Mazaruni Station, Sandwith 1548 (K, NY, U); Penal Settlement, Waby 8363 (BRG, NY). French Guiana: Trois-Sauts, Grenand 656 (CAY, P).

Note: The only differences with *A. paramaribensis* are mentioned in the key (14). In the future, with more collections available showing the variability of this species, *A. consimilis* might prove to be a synonym. In the treatment for Flora of Ecuador, González prefers to place *A. consimilis* in the synonymy of *A. acutifolia* Duchartre.

3. **Aristolochia cremersii** Poncy, Bull. Mus. Hist. Nat. (Paris), sér. 4, 10: 339, tab. 2. 1989 ("1988"). Type: French Guiana, Sommet Tabulaire, Cremers 6455 (holotype P, isotypes CAY, P, U, US).

Woody vine; stem to 6-7 cm diam., bark thick and fissured; branchlets glabrous; pseudostipules lacking. Leaves: petiole twining, up to 6 cm long; blade chartaceous, widely triangular, 9-14 x 7-14 cm, glabrous, shining above, greyish and slightly tomentose beneath, apex acute to acuminate, base truncate to slightly cordate. Inflorescence a cauliflorous or axillary raceme up to 10 cm long, bearing 10-15 flowers with very small (1 mm long) petiolate bracts; pedicels thin, up to 5 cm long. Flowers: ovary 8-10 mm long; perianth glabrous; utricle globose, 3-5 mm diam., with a small, scaly corona at base; tube 2-2,5 cm long, yellowish, with brown veins; limb with superior lobe brown, sublinear, 3-3.5 x 0.2-0.4 cm, inferior lobe 0.5-0.8 cm, gutter-shaped, emarginate. Infructescence up to 20 cm long with 4 mm long bracts; fruit glabrous, 3-5 cm long; seeds ovoid, 5 x 3 mm, acuminate at apex, emarginate at base.

Distribution: Known only from central French Guiana (FG: 3).

Specimens examined: French Guiana: Paul Isnard, Feuillet 308 (CAY, US); Arataye R., Saut-Pararé, Poncy 217 (CAY, P).

4. **Aristolochia daemoninoxia** Masters, Bull. Misc. Inform. 1906: 6. 1906. Type: Guyana, Demerara, Jenman 6915 (holotype K, isotypes BRG, K, NY). – Plate 3. 6-10.

Woody vine; stem 2 cm diam., bark deeply ridged, fissured; branchlets glabrous; pseudostipules lacking. Leaves: petiole twining, up to 7 cm long; blade chartaceous, triangular, 8-15 x 8-18 cm, glabrous, shining above, glaucous, velvet-tomentose beneath, apex obtuse or acute (acuminate in Gillespie 830), base straight to slightly cordate; lateral veins marginal at base. Inflorescence cauliflorous, or rarely axillary (Fanshawe 1383); racemes contracted (up to 1 cm, mostly less than 5 mm), bearing few flowers with minute scaly bracts; pedicels 2-3.5 cm long. Flowers (poorly known): ovary 1-1.5 cm long; perianth tomentose outside when young, pale brownish green with darker veins; utricle ovoid, 2 x 1.3 cm; tube short, funnel-shaped; superior lobe ovate, wider than long, 3 x 5-6 cm, apex truncate and mucronate, inside verrucose. Fruit 7-10 cm long, rough, glabrous, midrib of carpels thickened, with a median groove, external wall thick, woody; seeds ovoid, 5 x 2.5 mm, apex rounded, base emarginate, with the funicle slightly winged, longitudinally protruding.

Distribution: Only known from Guyana (GU: 4).

Specimens examined: Guyana: Jenman 6543 (BRG); Mazaruni Station, Fanshawe 1383 = FD 4119 (K, NY); Kato, Gillespie 830 (BRG, P, US).

Vernacular names: Guyana: boehari, boohiari, boyarri (Arawak), devildoer (Creole), pauisima (Arawak).

5. **Aristolochia didyma** S. Moore, J. Bot. 53(1): 7 and 53(2): tab. 535, fig. 1. 1915. Hoehne, Fl. Brasílica 15(2): tab. 101. 1942. Type: Brazil, Pará, Moss s.n., in 1912 (holotype probably at BM (not seen), isotype US).

Aristolochia rodriguesii Hoehne (as "rodriguessi"), Mem. Inst. Oswaldo Cruz 20(1): 76 (140), tab. 72. 1927. Lectotype here designated: Bolivia, Ule 9338 (holotype MG).

Vine; branchlets glabrous, hollow; pseudostipules lacking. Leaves glabrous: petiole twining, 5-8 cm long; blade membranous, triangular, widely cordiform, 8-25 x 6-17 cm, glaucous beneath, apex triangular, acute to acuminate, base deeply cordate with acute sinus, margin undulate; lateral veins marginal at base. Flowers axillary, solitary; pedicel thin, 5-8 cm long (up to 11 cm in Black 47-2159, Pará, Brazil), glabrous; ovary ± 2 cm long (up to 6 cm in Black 47-2159); perianth yellow green outside, glabrous; utricle ovoid, 3-4 x 5-6 cm, black spotted; tube cylindric, about 3 x 1-1.5 cm; superior lobe with margin usually shortly violet-fimbriate, purple veined, 4-5 cm long, 2-fid, each lobule about 2.5 x 1 cm, apex rounded, inferior lobe 1.5-2 cm long, throat yellow. Fruit unknown.

Distribution: Colombia, Venezuela, French Guiana, Peru, Amazonian Brazil (+1 collection from Espirito Santo), Bolivia; in forests on slopes; 10 collections studied (FG: 4).

Selected specimens: French Guiana: Mts. de Kaw, Skog & Feuillet 5675 (CAY); Atachi Bacca, de Granville et al. 10947 (CAY, P, US); Saül, Cremers 6069 (CAY).

Note: In spite of its large, odd looking and rather showy flowers, and of its large distribution area, this species has been very poorly collected, probably because the axillary flowers are inaccessible on this large vine.

6. **Aristolochia flava** Poncy, Bull. Mus. Hist. Nat. (Paris), sér. 4, 10: 341, tab. 3, fig. 1. 1989 ("1988"). Type: French Guiana, Sinnamary R., Petit Saut, Prévost 1770 (holotype P, isotypes CAY, P).

Woody vine; stem 0.5-0.7 cm diam., with a relatively thin bark; branchlets glabrous; pseudostipules lacking. Leaves: petiole thin, 3-5 cm long; blade triangular, 11-17 x 9-12 cm, glabrous, shining above, greyish, pubescent beneath, apex long acuminate, acumen 2-2.5 cm long, base straight to slightly cordate; lateral veins submarginal at base. Inflorescence axillary to cauline, racemes thin, long (up to 25 cm), with a long peduncle, bearing 15-20 flowers with minute (< 1 mm long) petiolate bracts; pedicels very thin, 3 cm long. Flowers: ovary 1 cm long; perianth entirely yellow, glabrous; utricle globose, about 0.5 cm in diam.; tube slightly funnel-shaped, 1.5 cm long; limb funnel-shaped, inferior lobe widely elliptic, 2 x 3-3.5 cm, superior lobe linear, twisted, about 4 cm long, shortly papillose inside. Fruit unknown.

Distribution: Known only from the type collection from lowland forest in northern French Guiana.

7. **Aristolochia guianensis** Poncy, Bull. Mus. Hist. Nat. (Paris), sér. 4, 10: 343, tab. 3, fig. 2. 1989 ("1988"). Type: French Guiana, Mana R., Saut Sabbat, de Granville 4530 (holotype P, isotype CAY).

Herbaceous, white pubescent vine; pseudostipules lacking. Leaves: petiole thin, 3-6 cm long; blade ovate, 8-11(-12) x 5-7(-9) cm, apex acute, base deeply cordate; lateral veins marginal at base. Flowers axillary, solitary; pedicel thin, 5-8 cm long; ovary 1-1.5 cm long, more densely pubescent than the pedicel; perianth wine-red, yellowish green veined outside; utricle pyriform, 1.7 x 1.7 cm; tube cylindric, 1.5-2 cm long; limb funnel-shape, opening narrowed in the middle, superior lobe vertical, triangular, 2-3 cm wide at base, apex shortly acuminate, inside white, with dark violet spots distally. Fruit unknown.

Distribution: Known from the coastal area of E Suriname and W French Guiana; in open vegetation at the edge of the forest (FG); (SU: 2; FG: 5).

Selected specimens: Suriname: Brownsberg, Zaandam BW 6632 (U); Lely Mts., Lindeman, Stoffers et al. 617 (U). French Guiana: between Organabo and Saut Sabbat, Feuillet 2330 (CAY, P); near Saut Sabbat, Sastre & Bell 8031 (CAY, P, US).

8. **Aristolochia hians** Willdenow, Mém. Soc. Imp. Naturalistes Moscou 2: 100. 1809. – *Howardia hians* (Willdenow) Klotzsch, Monatsber. Königl. Preuss. Akad. Wiss. Berlin 1859: 607. 1860. Type: Venezuela, Bredemeyer 27 (holotype B-W).

Vine; branchlets hollow, glabrous; pseudostipules foliaceous, cordiform, up to 2.5 cm long. Leaves: petiole thin, 3-5 cm long; blade round, cordiform, 6-10 x 6-11 cm, glabrous, apex obtuse to rounded, base cordate; lateral veins marginal at base. Flowers axillary, solitary; pedicel twining, 7-8 cm long; ovary 3-4 cm long, bending at a right angle from the pedicel; perianth whitish outside, purple-veined, glabrous; utricle obpyriform, 2-6 x 1.5-4 cm, its very base purple; tube cylindric, up to 1.5 cm long; limb 2-lobed, rounded and hanging, 4-6 x 5-7 cm, pale brown, superior lobe paddle-shaped, linear and upright at base, 1.5-3 cm long, inferior lobe linear, 7-9 (up to 15) x 0.5-1.5 cm. Fruit 8 cm long, glabrous, chartaceous, carpels keeled along midrib; seeds triangular, 3-4 mm long and wide, winged, 13-16 x 7-10 mm including wing, slightly emarginate at base, very thin.

Distribution: Venezuela, Guyana, Brazil; in secondary forest and shrubby vegetation; 17 collections studied (GU: 5).

Specimens examined: Guyana: Kato, Gillespie 869 (P, US); Barima R., Jenman 6934 (BRG, K, NY), Im Thurn s.n. (K), Rob. Schomburgk I add. ser. 144 S (K); no loc., Jenman 7227 (BRG)?

9. **Aristolochia iquitensis** O.C. Schmidt, Notizbl. Bot. Gart. Berlin-Dahlem 10: 196. 1927. Type: Peru, Iquitos, Tessmann 5120 (holotype B, probably destroyed, photo COL, F, G, GH, MO, NY).

Aristolochia macrophylla Duchartre, Ann. Sci. Nat., sér. 4, 2: 68. 1854, non Lamarck 1783. – *A. magnifolia* Briquet, nom. inval.: ms. on the type. Type: French Guiana, Cayenne, Leprieur s.n., in 1839 (holotype G-DEL, photo F, US).

Woody vine; stem to ca. 1 cm in diam., bark relatively thin, irregularly thickened; branchlets glabrous; pseudostipules lacking. Leaves: petiole twining, 3-6 cm long; blade chartaceous, elongate elliptic with strongly curved to parallel margins, 8-17(-20) x 5-7 cm, glabrous above, barely pubescent beneath, apex usually acuminate (5-10 mm long), base deeply cordate to auriculate; lateral veins usually marginal at very base. Inflorescence cauliflorous; 1-2 racemes, 2-6 cm long, bearing commonly 10 flowers or many more; bracts small, scaly; pedicels thin, 3.5-5 cm long (in the Guianas, shorter in the type specimen). Flowers: ovary 1-1.5 cm long, thin, glabrous, pink; perianth glabrous, purple brown outside; utricle ovoid, 7 x 4.5 mm; tube funnel-shaped, short, usually less than 1 cm long; limb with 2 lateral lobes which are more or less triangular and oblique downward, variable in size, apex round, yellow with brown markings, lower margin fleshy fimbriate. Fruit 6-8 cm long, glabrous; seeds ovoid, 5 x 3 mm, apex acute, base emarginate.

Distribution: Amazonian regions of Colombia, Peru and Brazil, and the Guianas; in primary and secondary forest; 18 collections studied (GU: 1; SU: 1; FG: 8).

Selected specimens: Guyana: Pakaraima Mts., Maas et al. 4184 (U). Suriname, Maroni R., Hugh-Jones 32 (K, U). French Guiana: Mt. Nouragues, Prévost 2184 (CAY); Piste de St. Elie, Feuillet 2282 (CAY); Saül, de Granville 928 (CAY, P).

Use: Sticks boiled for stomach ache.

Vernacular name: Suriname: kwashi-bita.

Note: The colour and structure of the perianth are constant, but the size and orientation of the lobes, and the size of the flowers are variable.

10. **Aristolochia leprieurii** Duchartre (as "leprieurei"), in De Candolle, Prodr. 15(1): 451. 1864. – *Aristolochia rubromarginata* Fischer, nom. nud., on the holotype and the LE isotype. Type: French Guiana, Cayenne, Leprieur s.n. (holotype P, isotypes G-DEL, LE, photo F, NY, US). – Plate 1.

Aristolochia gabrielis Duchartre ex Briquet, Candollea 4: 351. 1931. Type: French Guiana, Gabriel s.n. (anno 1802) (holotype G-DEL, photo F, US, isotype F).

Sarmentose vine, with thick woody rootstock; branchlets thin, hirsute; pseudostipules lacking. Leaves: petiole thin, more densely hirsute, 1.5-3 cm long; blade chartaceous, triangular, cordiform, 6-12 x 3-6 cm, glabrous above, finely pubescent beneath, apex variable in one plant, base cordate; venation pedate, marginal at base. Flowers axillary, solitary; pedicel thin, pubescent like the petioles, 2-3.5 cm long; ovary more pubescent than pedicel, ± 1 cm long; perianth white, pubescent; utricle whitish green, ovoid, 8-15 x 5-8 mm, upper part with an area of smaller, probably secreting cells; tube dark purple, straight, up to 1.5 cm; limb funnel-shaped, superior lobe triangular, gutter-shaped, up to 2.5 cm long, apex acute, violet, marginally reflexed, fleshy fimbriate inside, throat pale yellow. Fruit about 4 cm long, sometimes with sparse hairs, rostrum up to 1 cm; seeds triangular, 4 x 4 mm, funicle slightly winged, longitudinally protruding.

Distribution: Venezuela (1 coll., Hahn 4915 (P, US) from Portuguesa), Guyana, French Guiana; rather common weed between Cayenne and Saint-Georges (FG); 17 collections studied (GU: 1; FG: 15).

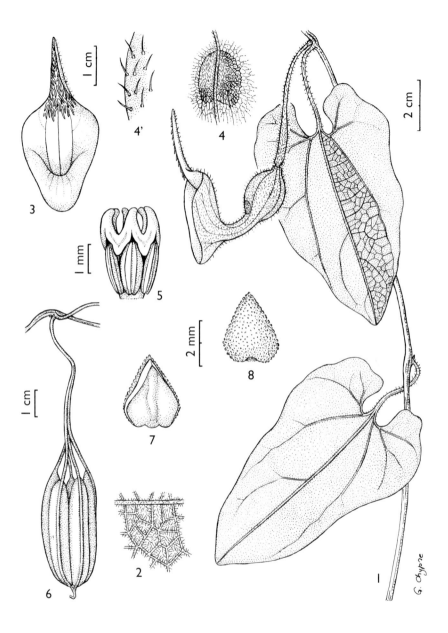

Plate 1. *Aristolochia leprieurii* Duchartre: 1, stem with leaves and flower; 2, detail of venation; 3, flower: frontal view of perianth limb; 4, detail of the area with secreting cells on the upper utricle; 4', detail of hairs on erect lobe of perianth; 5, gynostemium; 6, fruit; 7-8, seed, both faces (1-5, Poncy 2; 6-8, Poncy 3).

Selected specimens: Guyana: Essequibo R., near mouth of Blackwater Cr., A.C. Smith 2831 (P). French Guiana: Matoury, Feuillet 1373 (CAY, P), de Granville 682 (CAY, P).

Use: Entire plant boiled and used externally to reduce fever (from label, A.C. Smith 2831). Medicinally used by the Wayampi people against diarrhoea, also known by the Palikour people and the Creole population, according to Grenand et al. (1987) who treated this species together with *Aristolochia mossii* S. Moore.

Vernacular names: Guyana: mametala (Waiwai). French Guiana: bukuti (Palikour); liane amère (Creole); ulu?ay, uluwu?ay (Wayampi).

11. **Aristolochia mossii** S. Moore, J. Bot. 53(1): 7 and 53(2): tab. 535, fig. 2. 1915. Type: Brazil, Pará, Moss 86 (holotype BM? (not seen), isotype US).

Aristolochia dubia Hoehne, Mem. Inst. Oswaldo Cruz 20(1): 111(175), tab. 103. 1927. Type: Brazil, Pará, Moss MG 15482 (holotype MG).

Vine, stem glabrous; pseudostipules lacking. Leaves: petiole 2.5-6 cm long; blade papyraceous when dry, triangular-oblong, 6-15 x 3-6 cm, glabrous, shining above, paler beneath, apex acute, base deeply cordate; venation pedate. Flowers axillary, solitary; pedicel glabrous, 5-7 cm long; ovary 2.5-4 cm long, glabrous; perianth yellow outside, glabrous; utricle obovoid, 3 x 1.5 cm, veins and dots dark purple outside; tube large, 3 x 1 cm, with less purple; lobe unique, superior, more or less rectangular and small, 1 x 1 cm, partly closing the tube, with dark purple margin, apex emarginate, inside dark purple near apex and yellow at base, throat yellow. Fruit 6-7 x 2 cm, glabrous; seeds cordiform, 4 x 4 mm, wing obovate, 9-11 x 8-10 mm including wing.

Distribution: Venezuela, French Guiana, Brazil (Pará); 9 collections studied (FG: 6).

Specimens examined: French Guiana: Richard s.n. (P); Trois Sauts, Grenand 260 (CAY), 565 (CAY), Lescure 374 (CAY), Prévost & Grenand 1491 (CAY), 1941 (CAY, P).

Use: See use of *Aristolochia leprieurii* Duchartre.

Vernacular name: French Guiana: uluwu?ay (Wayampi).

12. **Aristolochia paramaribensis** Duchartre, in De Candolle, Prodr. 15(1): 496. 1864. Type: Suriname, near Paramaribo, Wullschlägel 452 (holotype in herb. Mart.). – Plate 2.

Vine, glabrescent, pubescent when young; branchlets angular to obtuse-angled; pseudostipules lacking. Leaves: petiole 1-2 cm; blade coriaceous, elliptic, 8-15 x 3-8 cm, shining above, apex emarginate or round or acute to acuminate, base narrowly cordate, sometimes with overlapping basal lobes; venation pedate, secondary veins parallel. Inflorescence axillary, 5-6 cm long, tomentose, bearing 7-10 flowers; bracts linear, 3-5 mm long; pedicels minute to absent. Flowers: ovary up to 1 cm long; perianth brownish green outside, tomentose when young, becoming sparsely tomentose along veins; utricle ovoid, 1 x 0.5-0.7 cm; tube about 1 cm; limb funnel-shaped, 3 x 1.5 cm, superior lobe triangular, 1-1.5 cm, apex acute with terminal filiform appendage. Fruit 6-7 cm long, glabrous, inner walls membranaceous, comb-like; seeds cordiform, 5 x 5 mm, bearing 2 parallel, rectangular, papyraceous wings, ca. 10 x 15 mm including wings, base truncate.

Distribution: The Guianas, Brazil, Peru; in secondary vegetation in the coastal area; 10 collections studied (GU: 1; SU: 5; FG: 4).

Selected specimens: Guyana: Essequibo Islands, Gillespie 1018 (BRG, P, US). Suriname: near Paramaribo, Heyde 401 (P, U); Albina, Went 425 (U). French Guiana: St. Jean du Maroni, Prévost 621 (CAY, P); Maroni, Mélinon 381 (P).

Note: An anonymous specimen is annotated *Aristolochia guyanensis*. This name has not been published. The specimen (Anonymous 44) belongs to *Aristolochia paramaribensis* Duchartre and not to *Aristolochia guianensis* Poncy.

13. **Aristolochia peltatodeltoidea** Hoehne, Fl. Brasílica 15, 2(6): 102, tab. 75. 1942. (as "peltato-deltoidea"). Type: Guyana, Kanuku Mts., A.C. Smith 3385 (holotype US, isotypes F, G, K, MO, NY, P, U).

Glabrous vine with woody rootstock; branchlets sarmentose, thin; pseudostipules lacking. Leaves: petiole thin, 1-2(-3) cm long; blade triangular, 3-7(-10) x 1.5-3(-5.5) cm, greyish and pubescent beneath, apex triangular acute, base straight to slightly cordate, 1-2 mm peltate; venation pedate, tertiary veins forming an inconspicuous net. Flowers axillary, solitary, or thin racemes up to 2 cm long, bearing up to 6 flowers with small linear bracts; pedicels very thin, 5-8 mm long; ovary 3-5 mm

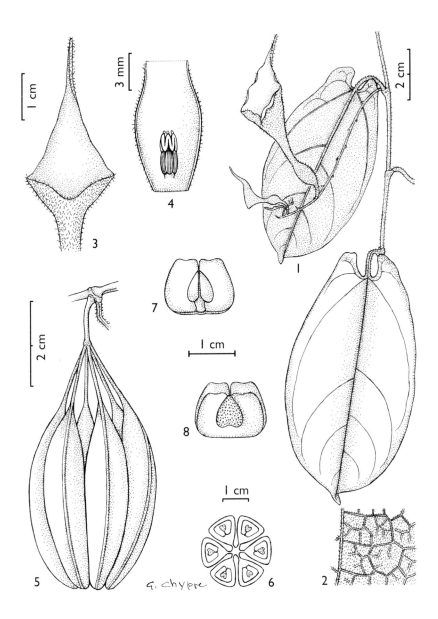

Plate 2. *Aristolochia paramaribensis* Duchartre: 1, leaves and inflorescence; 2, detail of reticulate venation; 3, upper perianth; 4, gynostemium in the utricle; 5, fruit; 6, transverse section of young fruit; 7-8, seed, both faces (1-4, Prévost 621; 5-8, Aubréville 89).

long; perianth pale green, glabrous; utricle subglobose to ovoid, 4-6 mm in diam.; tube cylindric, 3 mm; limb funnel-shaped, superior lobe large, brown, elliptic, 1.5 x 1 cm, apex round. Fruit unknown.

Distribution: Only known from Guyana, Kanuku Mts., on rock boulders (GU: 3).

Specimens examined: Guyana: S of Kanuku Mts., Maas et al. 4058 (U); Kanuku Mts., Nappi Cr., Jansen-Jacobs et al. 705 (BRG, P, U); Kanuku Mts., foothills at Moco Moco R., rocky outcrop, Jansen-Jacobs et al. 4592 (BRG, P, U).

14. **Aristolochia rugosa** Lamarck, Encycl. 1: 252. 1783. – *Aristolochia obtusata* Swartz, Prod. Veg. Ind. Occ. 126. 1788. – *Howardia obtusata* (Swartz) Klotzsch, Monatsber. Königl. Preuss. Akad. Wiss. Berlin 1859: 612. 1859. Type: West Indies, see Plumier ic. 27, t. 33.

Aristolochia barbata Jacquin, Ic. Pl. Rar. 3, 6: 608. 1789. – *Howardia barbata* (Jacquin) Klotzsch, Monatsber. Königl. Preuss. Akad. Wiss. Berlin 1859: 613. 1859. Type: Venezuela (holotype W (not seen)). *Howardia schomburgkii* Klotzsch, Monatsber. Königl. Preuss. Akad. Wiss. Berlin 1859: 613. 1859. – *Aristolochia rumicifolia* Rich. Schomburgk (not Martius et Zucc.), in sched. (414), nom. nud. – *Aristolochia dictyantha* Duchartre var. *schomburgkii* (Klotzsch) Duchartre, in De Candolle, Prodr. 15(1): 447. 1864. – *Aristolochia rumicifolia* Schomburgk ex Duchartre, loc. cit., pro syn., nom. nud. Type: Guyana, Pirara, Rich. Schomburgk 611 (holotype B not extant). Neotype proposed here: Guyana, Pirara, Rich. Schomburgk II 414 (holotype P, isotype BM, K).

Vine with woody rootstock; stem sarmentose; branchlets glabrous; pseudostipules lacking. Leaves: petiole 2-3 cm long; blade triangular to sagittate, 6-13 x 4-8 cm, glabrous, shining above, greyish beneath, apex acute to rounded, base slightly to strongly cordate; venation pedate. Inflorescence axillary or flower solitary; pedicels 2-5 cm long. Flowers: ovary 0.7-15 mm long; perianth pale brown with yellow veins outside, glabrous; utricle ovoid, 10-13 x 6-8 mm; tube cylindric, 1-1.5 cm; limb funnel-shaped, about 1.5 in diam., lobe unique, superior, erect, 1-1.5 cm long, yellow at base, distal part violet brown, with fleshy, dark purple fimbriae, ovoid, apex acuminate. Fruit (from Caribbean collections) 3-3.5 cm long, glabrous; seeds 3-4 mm long.

Distribution: Widely distributed in the Caribbean region, W to Brazil and S to Bolivia; in open vegetation; more than 30 collections studied (GU: 4; SU: 1; FG: 2).

S p e c i m e n s e x a m i n e d : Guyana: without locality, Schomburgk 114 (BM, K), Rob. Schomburgk 85 (K, P); Karasabaï, Feuillet 10642 (US). Suriname: Upper Coppename R., Boon 1149 (U). French Guiana: without locality, Benoist 891 (P); Saint Laurent, Mélinon 428 (P).

N o t e : A recent collection (Cremers and Crozier 14677), that came in after this treatment was completed, does not match with any of the species treated. It could belong to *A. disticha* Masters which had not yet been recorded from the Guianas area (F. González, pers. comm.). F. González, who is carrying out a monographic study of the genus *Aristolochia*, also has informed us that two other specimens from the Guianas (Benoist 891 and Melinon 428) cited in this treatment under *A. rugosa* would fit better in *A. disticha*. Both species are very close, but the surface of the leaf of *A. disticha* is smooth and the flowers are arranged in short distichous racemes, while *A. rugosa* has rugose leaves and axillary flowers. The authors express their acknowledgments to F. González for providing this information.

15. **Aristolochia stahelii** O.C. Schmidt (as "staheli"), Feddes Repert. Spec. Nov. Regni Veg. 45: 52. 1938. Type: Suriname, Brownsberg Top, Stahel & Gonggrijp BW 6520. (holotype U, isotype NY).

– Plate 3. 1-5.

Woody vine; stem with corky expansions, 1-3 cm in diam., 6-7 mm without bark; branchlets glabrous; pseudostipules lacking. Leaves: petiole 5-7 cm long; blade chartaceous, widely cordiform, 7-13(-18) x 10-15(-22) cm, glabrous and shining above, greyish and pubescent beneath, apex obtuse or rounded to acuminate, base slightly cordate; venation pedate, marginate at base. Inflorescence cauliflorous; racemes short (up to 1 cm long), bearing 4-5 flowers with minute scaly bracts; pedicels 2-4 cm long. Flowers: ovary about 2-3 cm long; perianth whitish pink with reddish veins outside, glabrous; utricle ovoid, 3 x 1.5 cm; tube reduced and hidden by the limb; limb red purple, not spreading, but at first reflexed backwards to the utricle and apically coming forward, finally recurved outwards and forming a horse-shoe-shaped cavity around the mouth of the tube, apex acute, reflexed backward. Fruit 8-10 cm long, glabrous, midrib of carpels thickened, external wall thick, woody; seeds triangular, 6 x 3 mm, funicle longitudinally protruding.

D i s t r i b u t i o n : Suriname, French Guiana, Brazil (Amapá); in primary and old secondary forests; 8 collections studied (SU: 3; FG: 4). Many sterile collections from French Guiana probably belong to this common species.

18

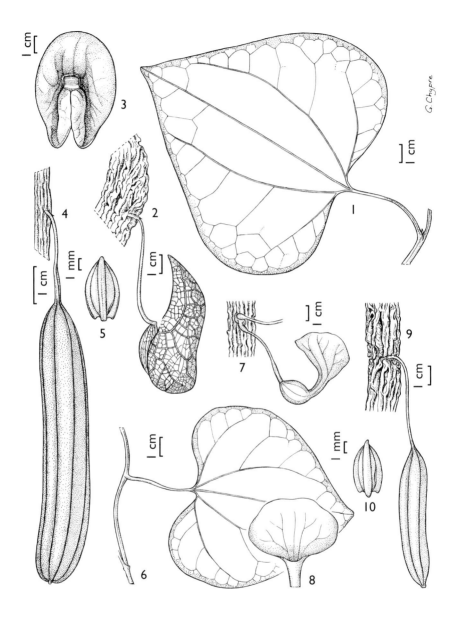

Plate 3. 1-5, *Aristolochia stahelii* O.C. Schmidt: 1, leaf; 2, stem and flower in profile view; 3, perianth pavilion in front view; 4, fruit; 5, seed (1,4,5, BW 3281; 2-3, Stahel & Gonggrijp 6520). 6-10, *Aristolochia daemoninoxia* Masters: 6, leaf; 7, stem and flower in profile view; 8, perianth pavilion in front view; 9, fruit; 10, seed (6-10, Jenman 6915).

Specimens examined: Suriname: Lely Mts., Lindeman, Stoffers et al. 370 (U); Brownsberg, BW 3281 (NY, U). French Guiana: Piste de St. Elie, Prévost 2154 (CAY, P); Mts. de Kaw, Feuillet 2103 (B, BR, CAY, MO, NY, P, U, US); Mts. de l'Observatoire, de Granville 6752 (CAY, P).

Use: Medicinally used in French Guiana, mainly by the Creole population near Saint-Georges to stop fever or as a remedy against diabetes, also used by the Palikour people.

Vernacular names: French Guiana: bukuti (Palikour); liane amère (Creole); ulu?ay, uluwu?ay (Wayampi).

16. **Aristolochia surinamensis** Willdenow, Sp. Pl. 4(1): 151. 1805. – *Aristolochia trilobata* Linnaeus sensu Lamarck, Encycl. 1: 251. 1783, non Linnaeus. – *Howardia surinamensis* (Willdenow) Klotzsch, Monatsber. Königl. Preuss. Akad. Wiss. Berlin 1859: 613. 1860. Type: Suriname, Herb. N.J. Jacquin (not seen).

Aristolochia platyloba Garcke, Linnaea 22: 69. 1849. Type: Suriname, N of Paramaribo, Tourtonne Plantation, Kegel 632 (holotype, GOET), syn. nov. *Aristolochia macrota* Duchartre, Ann. Sci. Nat., sér. 4, 2: 43. 1854. Lectotype here designated: Rob. Schomburgk I 679 (holotype G-DC, photo F, US, isotypes BM, F, K (2 sheets), US), syn. nov.

Vine; branchlets glabrous; pseudostipules lacking. Leaves: petiole thin, 2.5-5 cm long; blade 3-lobed, 5-9 x 7-11 cm, glabrous, shining above, greyish with very fine indument beneath, apex rounded, base truncate, sometimes slightly cordate, lateral lobes perpendicular to the middle lobe, middle lobe 3-6 cm long; venation pedate, marginal at base. Flowers axillary, solitary; pedicel 2.5-4 cm long; ovary about 1.5 cm long; perianth sparsely pilose to glabrous; utricle ovoid, 1.0-1.5 x 0.8-1.2 cm; tube about 1.5 cm long; limb funnel-shaped, superior lobe rounded, triangular, up to 3.5 x 3 cm, apex emarginate. Fruit small, 3.5 cm long, acumen 5 mm long, pericarp membranous, glabrous; seeds 2 mm long.

Distribution: Guyana and Suriname; 10 collections studied (GU: 2; SU: 8).

Selected specimens: Guyana: Rob. Schomburgk I 679 (BM, G, F, L, P); Rupununi distr., Acevedo 3308 (US). Suriname: Hostmann 611 (BM, GH, P, U); Lelydorp, Stahel 150 (NY, U).

Use: Febrifuge.

Vernacular name: Suriname: loango tete.

17. **Aristolochia trilobata** Linnaeus, Sp. Pl. 2: 960. 1753. – *Aristolochia triloba* Salisbury, Prodromus 1796: 214. 1796. Type: South America, Herb. Linnaeus 1071-1 (holotype LINN).

Aristolochia macroura Gomez, Mem. Acad. Real Sci. Lisboa 2: 29-34, t. 4. 1803. Type: Brazil, Martius 1817 (not seen).

Vine; branchlets glabrous; pseudostipules leafy, rounded, about 1.5 cm in diam. Leaves: petiole glabrous, 1.5-3 cm long; blade 3-lobed, 5-9 x 6-10 cm, glabrous, shining above, pubescent beneath, lobes very variable in shape and size, middle one often diamond-shaped, lateral lobes often falciform, always pointing forward, apex rounded, base straight; venation pedate, submarginal at base. Flower axillary, solitary; pedicel about 4 cm long; ovary 2 cm long; perianth greenish yellow, dark red veined outside, glabrous; utricle ovoid, 4-5 x 4 cm, with basal, retrorse appendages, 3-4 mm long; tube cylindric, wide from the base, 4-5 x 1.5 cm; superior lobe ovoid, with a long apical filament of 10(+) cm long, inside whitish. Fruit 6 cm long, glabrous; seeds flat, as wide as long, ca. 6-7 mm.

Distribution: Caribbean Islands and from Honduras to Argentina; common vine in open vegetation, especially near villages, where it is often protected if not cultivated; 17 collections studied (GU: 8; SU: 1; FG: 3).

Selected specimens: Guyana: Im Thurn 1945 (K); Essequibo, Wakenaam, Jenman s.n. (NY). Suriname: Wia Wia Bank, Lanjouw & Lindeman 1069 (U). French Guiana: Poiteau s.n. (K); Piste de St. Elie, Sastre 6897 (P); Maroni, Mélinon 236 (P).

Use: Cultivated in French Guiana as a medicinal plant (Moretti 104, 993) by the Creole population: leaves (or the caterpillars of *Parides* feeding on the flowers) are used to cure mosquito and snake bites, liver troubles, malaria.

Vernacular names: French Guiana: feuille trèfle, liane trèfle, trèfle (Creole); trèfle caraïbe (Caribbean Creole).

18. **Aristolochia weddellii** Duchartre, Ann. Sci. Nat., sér. 2, 4: 62. 1854. Type: Brazil, Matto Grosso, Weddell 3396 (holotype P).

In the Guianas only: var. **rondoniana** (Hoehne) Hoehne, Mem. Inst. Oswaldo Cruz 20(1): 59 (123), tab. 50. 1927. – *Aristolochia weddellii* Duchartre subsp. *rondoniana* Hoehne, Arch. Bot. Est. São Paulo 1(1): 19, tab. 8. 1925. Type: Brazil, Comm. Rondon 2445 (lectotype here designated SP, isotype RB).

Aristolochia longecaudata Masters, in Martius, Fl. Bras. 4(2): 84. 1875. Type: Guyana, Appun 288 (holotype K).

Vine; branchlets pubescent when young, glabrescent; pseudostipules foliaceous, orbicular. Leaves: petiole pubescent, 2-5 cm long; blade rounded to triangular, (5-)12-15 x (3.5-)6-12 cm, glabrous above, tomentose beneath, apex triangular-acute, base widely rectangular-cordate; venation pedate, marginal at base. Flowers axillary, solitary; pedicel thick, 7-12 cm long; ovary 3-5 cm long; perianth whitish, dark-purple veined outside; utricle with a basal thickened rim, asymmetric, ventricose, (3.5-)5-7 x (2-)3-4 cm; tube short to inconspicuous; limb short funnel-shaped, superior lobe long tape-like, scalloped, about 8-10 cm long and more, with a terminal filament more than 10 cm long. Fruit unknown.

Distribution: The Guianas, Brazil; 14 collections studied (GU: 6; SU: 1; FG: 3).

Selected specimens: Guyana: Mazaruni Station, Sandwith 1594 (B, NY, U); Persaud 33 = FD 5342 (NY). Suriname: Waremapan Cr., Stahel 70 (NY, U). French Guiana: Itany R., Aubert de la Rüe s.n. (P); Fleury 537 (CAY); de Granville et al. 10000 (CAY, P, US).

Note: Two other varieties, var. *weddellii* and var. *duckeana* Hoehne have the superior lobe of the perianth limb much longer; they are restricted to Amazonian Brazil.

DOUBTFUL SPECIES

Aristolochia amara (Aublet) Poncy, Bull. Mus. Hist. Nat. (Paris), sér. 4, 10: 344. 1989 "1988". – *Abuta amara* Aublet, Hist. Pl. Guiane 1: 620, tab. 251. 1775. Type: French Guiana, Aublet s.n. (holotype P, isotype BM).

Described from sterile material, this taxon cannot be identified with certainty with one of the above species. It could be any one of *A. bukuti*, *A. cremersii*, *A. daemoninoxia*, *A. stahelii* or a different species.

Aristolochia cymbifera Martius & Zuccharini, Nov. Gen. Spec. 1: 76, tab. 49. 1824. Type: Brazil, State of São Paulo, Martius s.n. (herb. Martius, not seen).

Vine, glabrous. Pseudostipules and leaves similar to *A. hians*. Flowers axillary, solitary; pedicel twining, 7-8 cm long; ovary 3-4 mm long, perpendicular to the pedicel; hypanthium present; perianth: utricle 5-6 x

2-3 cm, tube cylindric, up to 1.5 cm, superior lobe paddle-shaped, up to 12 cm long, 4 cm wide in the distal part, inferior lobe not seen in the Guianan material. Fruit not seen.

Note: Two old specimens of this species are preserved at P. One, in Herb. Richard, might have been collected in French Guiana by Leprieur, the other from Herb. Mocquin-Tandon seems to be of horticultural origin. The presence of this species, originating from southern Brazil (2 specimens studied), in the wild in the Guianas is very doubtful.

REJECTED SPECIES

Aristolochia bilobata Linnaeus, Sp. Pl. 960. 1753.

This species, cited by Aublet (1775, p. 833), is known only from Hispaniola and St. Thomas Islands. The leaves are 2-lobed, very typical, and there is no specimen from the Guianas that we could refer to that species. There are at least two possible explanations: either *Aristolochia bilobata* was cultivated in French Guiana or the name has been wrongly applied to one of the few 2-lobed species of *Passiflora* subgenus *Decaloba* occuring in French Guiana, whose young plants (up to 10-15 cm tall) have very minute stipules and no tendril. Even the second hypothesis is not more unlikely than the description of *Aristolochia amara* as a Menispermaceae.

Aristolochia brasiliensis Martius & Zuccharini, Nov. Gen. Sp. 1: 77. 1824.

This species, cited by Richard Schomburgk (3: 822. 1848), has a large distribution area through South America.

Aristolochia odoratissima Linnaeus, Sp. Pl. ed. 2: 1362. 1763.

This species, cited by Aublet (1775, p. 833) and by Richard Schomburgk (3: 822. 1848), has a large distribution area from the Caribbean and Mexico to Paraguay, but does not occur in the Guianas, eastern Venezuela, eastern Colombia and northern Brazil. What Aublet saw is quite obscure. According to Schomburgk, *Aristolochia odoratissima* was cultivated in Guyana and might have belonged, under Aublet's care, to the collections of the "Jardin du Roy" in Cayenne as well. See also Briquet (Candollea 4: 352. 1931) about a possible confusion with *Aristolochia gabrielis* (= *Aristolochia leprieurii*, 1-10).

Aristolochia peltata Linnaeus, Sp. Pl. 960. 1753.

This species, cited by Aublet (1775, p. 833) and by Richard Schomburgk (3: 822. 1848), is known only from eastern Cuba, Hispaniola and St. Thomas Islands. Schomburgk cited it from the savanna region, Guyana. But we found no specimen from the Guianas referable to *Aristolochia peltata*, whose 2-lobed leaves are very typical. Same comment as under *Aristolochia bilobata*.

WOOD AND TIMBER

by

BEN J.H.TER WELLE[3] & PIERRE DÉTIENNE[4]

WOOD ANATOMY

GENERIC DESCRIPTION

ARISTOLOCHIA Linnaeus – Fig. 1.

Growth rings absent.
Vessels diffuse, solitary (up to 90%), and some irregular clusters of 2-3, round to slightly oval, the longest axis is in tangential direction, 16-19(8-25) per sq. mm, diameter variable, in two distinct size groups, 50-55(35-85) and 150-190(115-210) µm, respectively. Vessel-member length: 256(184-311) µm. Perforations simple. Intervascular pits scarce, alternate, round to oval, 4-7 µm. Vessel-ray and vessel-parenchyma pits similar to slightly larger than the intervascular pits. Thin-walled tyloses scarcely present.
Rays large, interfascicular, dissecting the stem into bundles with fibre tracheids, vessels and parenchyma; multiseriate, 10-20 cells wide, 1(0-2) per mm, up to >9000 µm (= >220 cells) high. Heterogeneous, composed of various shapes of procumbent cells and few square/upright cells. Rhombic crystals abundant.
Parenchyma few, scanty paratracheal and some diffuse strands.
Fibre tracheids, thin-walled, lumen up to 12-18 µm, walls up to 3-4 µm. Pits clearly bordered, abundant on radial and tangential walls, 4-6 µm. Length: 914(690-1208) µm. F/V ratio: 3.57.

Studied: *A. daemoninoxia, A. stahelii.*

[3] Herbarium Division, Department of Plant Ecology and Evolutionary Biology, Heidelberglaan 2, 3584 CS Utrecht, The Netherlands.

[4] C.I.R.A.D.-Forêt, Maison de la Technologie, BP 5035, Montpellier, Cedex 1, France.

LITERATURE ON WOOD AND TIMBER

Carlquist, S. 1993. Wood and bark anatomy of Aristolochiaceae; systematic and habital correlations. IAWA Journl. 14(4): 341-357.

Metcalfe, C.R. & L. Chalk. 1950. Anatomy of the Dicotyledons. 2 Vols. Clarendon Press, Oxford.

Record, S.J. & R.W. Hess. 1943. Timbers of the New World. Yale University Press, New Haven, CT.

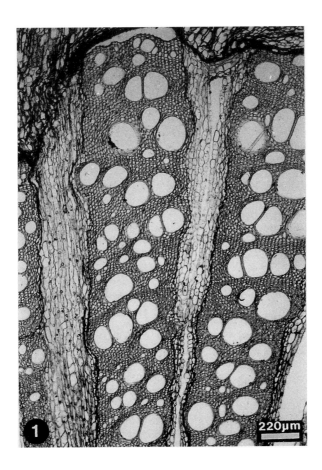

Fig. 1. Transverse section. *Aristolochia stahelii* O.C. Schmidt

NOMENCLATURAL CHANGES AND NEW TYPIFICATIONS

New synonyms:

Aristolochia platyloba Garcke, Linnaea 22: 69. 1849 = Aristolochia surinamensis Willdenow
Aristolochia macrota Duchartre, Ann. Sci. Nat., sér. 4, 2: 43. 1854 = Aristolochia surinamensis Willdenow

Typifications:

Aristolochia macrota Duchartre, Ann. Sci. Nat., sér. 4, 2: 43. 1854. Lectotype here designated: Rob. Schomburgk I 679 (lectotype G-DC, photo, F, US, isotypes BM, F, K (2 sheets), US).

Aristolochia rodriguesii Hoehne, Mem. Inst. Oswaldo Cruz 20(1): 76 (140), tab. 72. 1927. Lectotype here designated: Brazil, Ule 9338 (lectotype MG).

Howardia schomburgkii Klotzsch, Monatsber. Königl. Preuss. Akad. Wiss. Berlin 1859: 613. 1859. Neotype here proposed: Guyana, Pirara, Rich. Schomburgk II 414 (holotype P, isotype BM, K).

Aristolochia weddellii Duchartre subsp. *rondoniana* Hoehne, Arch. Bot. Est. São Paulo 1(1): 19, tab. 8. 1925. Lectotype here designated: Brazil, Comm. Rondon 2445 (lectotype SP, isotype RB).

NUMERICAL LIST OF ACCEPTED TAXA

1. Aristolochia Linnaeus
   - 1-1. A. bukuti Poncy
   - 1-2. A. consimilis Masters
   - 1-3. A. cremersii Poncy
   - 1-4. A. daemoninoxia Masters
   - 1-5. A. didyma S. Moore
   - 1-6. A. flava Poncy
   - 1-7. A. guianensis Poncy
   - 1-8. A. hians Willdenow
   - 1-9. A. iquitensis O.C. Schmidt
   - 1-10. A. leprieurii Duchartre
   - 1-11. A. mossii S. Moore
   - 1-12. A. paramaribensis Duchartre
   - 1-13. A. peltatodeltoidea Hoehne
   - 1-14. A. rugosa Lamarck
   - 1-15. A. stahelii O.C. Schmidt
   - 1-16. A. surinamensis Willdenow
   - 1-17. A. trilobata Linnaeus
   - 1-18. A. weddellii Duchartre var. rondoniana (Hoehne) Hoehne

COLLECTIONS STUDIED

Numerous collections without flower, and therefore impossible to identify, are not cited here. Numbers in italics represent type collections.

GUYANA

Acevedo, P., 3308 (1-16)
Appun, C.F., *288* (1-18)
Bartlett, A.W., s.n. (anno 1908) (1-2)
Fanshawe, D.B., 1181 = FD 3917 (1-2); 1383 = FD 4119 (1-4)
Feuillet, C., 10642 (1-14)
Gillespie, L.J. et al., 830 (1-4); 869 (1-8); 1018 (1-12)
Graham, V., 220 (1-2)
Hancock, J., s.n. (1-17)
Im Thurn, E.F., 1945 (1-17); s.n. (1-2); s.n. (1-8)
Jansen-Jacobs, M.J. et al., 705 (1-13); 4592 (1-13)
Jenman, G.S., 1945 (1-17); 4192 (1-18); 4615 (1-17); 6543, *6915* (1-4); *6916* (1-2); 6934 (1-8); 7226 (1-17); 7227 (1-8); s.n. (1-17); s.n. (1-18)
Maas, P.J.M. et al., 4058 (1-13); 4184 (1-9)
Persaud, C.A., 33 = FD 5342, 121 = FD 6772 (1-18)
Sandwith, N.Y., 1548 (1-2); 1594 (1-18)
Schomburgk, Rich. or Rob.?, 114 (1-14)
Schomburgk, Rich., *II 414* (1-14); *611*, not extant (1-14)
Schomburgk, Rob., *I 679* (1-16); I add. ser. 144 S (1-8); II 4 (1-17); 85 (1-14)
Smith, A.C., 2831 (1-10); *3385* (1-13)
Tutin, T.G., 136 (1-18)
Waby, J.F., 8363 (1-2)

SURINAME

Anonymous, 44 (1-12)
Boldingh, I., 3817B, s.n. (1-16)
Boon, H., 1149 (1-16)
BW, 3281, *6520* (1-15); 6632 (1-7)
Florschütz, J. & P.A., 893 (1-12)
Focke, H.C., s.n. (1-16)
Geijskes, D.C., 70 (1-18)
Heyde, N.M., 401 (1-12)
Herb. N.J. Jacquin, *s.n.* (1-16)
Hostmann, F.W., 611, s.n. (1-16)
Hugh-Jones, D.H., 32 (1-9)
Kappler, A., 1604 (1-16)
Kegel, H.A.H., *632* (1-16)
Lanjouw, J. & J.C. Lindeman, 1069 (1-17)
Lindeman, J.C., A.L. Stoffers et al., 370 (1-15); 617 (1-7)
Splitgerber, F.L., s.n. (1-16)
Stahel, G., 70 (1-18); 150 (1-16)
Stahel, G. & J.W. Gonggrijp, *BW 6520* (1-15)
Vreden, F.E., LBB 11303 (1-12)
Went, F.A.F.C., 425 (1-12)
Wullschlägel, H.R., *452* (1-12)
Zaandam, C.J., BW 6632 (1-7)

FRENCH GUIANA

Anonymous, s.n. (1-10)
Aubert de la Rüe, M., s.n. (1-18)
Aublet, J.B.C.F. d', *s.n.* (after 1-18)
Aubréville, A., 89 (1-12)
Benoist, R., 357 (1-7); 891 (see note 1-14); 953 (1-12)

Cremers, G., 6069 (1-5); *6455* (1-3)

Cremers, G. & F. Crozier, 14677 (see note 1-14)

Feuillet, C., 308 (1-3); 707 (1-16); 1373, 1565 (1-10); 2103 (1-15); 2282 (1-9); 2330 (1-7); 3004 (1-15)

Fleury, M., 537 (1-18)

Foresta, H. de, 548, 568 (1-15)

Gabriel, A., *s.n.* (1-10)

Granville, J.J. de (et al.), 682 (1-10); 928 (1-9); 3440 (1-17); *4530* (1-7); 6752 (1-15); 7968 (1-1); 10000 (1-18); 10300 (1-15); 10947 (1-5); 11703 (1-17); 13173 (1-9)

Grenand, P., 260, 565 (1-11); 656 (1-2)

Grenand, P. & M.F. Prévost, 2049 (1-10)

Herb. Rudge, s.n. (1-10)

Jacquemin, H., 1712 (1-11); 2180 (1-10); 2605 (1-17)

Leprieur, F.R.M., *s.n. (anno 1839)* (1-9); *s.n.* (1-10)

Lescure, J.P., 374 (1-11)

Loubry, D., 2925 (1-9)

Mélinon, E., 236 (1-17); 381 (1-12); 428 (see note 1-14)

Moretti, C., 104, 993 (1-17); 1147 (1-15)

Mori, S.A. et al., 21637, 22251 (1-1); 22953 (1-3); 22965, 23347, 23966 (1-1)

Oldeman, R.A.A., 2622 (1-1)

Poiteau, P.A., s.n. (1-17)

Poncy, O., 2, 3 (1-10); 217 (1-3); *536* (1-1); 987 (1-9)

Prévost, M.F., 604 (1-1); 621 (1-12); 1113 (1-15); 1353, 1485, 1753 (1-10); 1768 (1-15); *1770* (1-6); 2154 (1-15); 2184 (1-9)

Prévost, M.F. & P. Grenand, 1491, 1941 (1-11)

Richard, L.C., s.n. (1-11); s.n. (1-10)

Riéra, B., 273 (1-15)

Rohr, J.P.B. von, s.n. (1-10)

Sagot, P., 1159 (1-7); s.n. (1-9); s.n. (1-10)

Sastre, C., 5698 (1-12); 6897 (1-17)

Sastre, C. & D. Bell, 8031 (1-7)

Skog, L.E., & C. Feuillet, 5675 (1-5)

INDEX TO SYNONYMS, TYPE SPECIES, DOUBTFUL AND
REJECTED SPECIES, AND NAMES IN NOTES

INDEX TO VERNACULAR NAMES

Alphabetic list of families of series A occurring in the Guianas

Defined as in Cronquist, 1981, and numbered in his sequence, with alternative names. Those published, with chronological fascicle number and year.

Abolbodaceae			Callitrichaceae	150		
(see Xyridaceae	182)	15. 1994	Campanulaceae	162		
Acanthaceae	156		(incl. Lobeliaceae)			
(incl. Thunbergiaceae)			Cannaceae	195	1. 1985	
(excl. Mendonciaceae	159)		Canellaceae	004		
Achatocarpaceae	028		Capparaceae	067		
Agavaceae	202		Caprifoliaceae	164		
Aizoaceae	030		Caricaceae	063		
(excl. Molluginaceae	036)		Caryocaraceae	042		
Alismataceae	168		Caryophyllaceae	037		
Amaranthaceae	033		Casuarinaceae	026	11. 1992	
Amaryllidaceae			Cecropiaceae	022	11. 1992	
(see Liliaceae	199)		Celastraceae	109		
Anacardiaceae	129	19. 1997	Ceratophyllaceae	014		
Anisophylleaceae	082		Chenopodiaceae	032		
Annonaceae	002		Chloranthaceae	008		
Apiaceae	137		Chrysobalanaceae	085	2. 1986	
Apocynaceae	140		Clethraceae	072		
Aquifoliaceae	111		Clusiaceae	047		
Araceae	178		(incl. Hypericaceae)			
Araliaceae	136		Cochlospermaceae			
Arecaceae	175		(see Bixaceae	059)		
Aristolochiaceae	010	20. 1997	Combretaceae	100		
Asclepiadaceae	141		Commelinaceae	180		
Asteraceae	166		Compositae			
Avicenniaceae			(= Asteraceae	166)		
(see Verbenaceae	148)	4. 1988	Connaraceae	081		
Balanophoraceae	107	14. 1993	Convolvulaceae	143		
Basellaceae	035		(excl. Cuscutaceae	144)		
Bataceae	070		Costaceae	194	1. 1985	
Begoniaceae	065		Crassulaceae	083		
Berberidaceae	016		Cruciferae			
Bignoniaceae	158		(= Brassicaceae	068)		
Bixaceae	059		Cucurbitaceae	064		
(incl. Cochlospermaceae)			Cunoniaceae	081a		
Bombacaceae	051		Cuscutaceae	144		
Bonnetiaceae			Cycadaceae	208	9. 1991	
(see Theaceae	043)		Cyclanthaceae	176		
Boraginaceae	147		Cyperaceae	186		
Brassicaceae	068		Cyrillaceae	071		
Bromeliaceae	189	p.p. 3. 1987	Dichapetalaceae	113		
Burmanniaceae	206	6. 1989	Dilleniaceae	040		
Burseraceae	128		Dioscoreaceae	205		
Butomaceae			Dipterocarpaceae	041a	17. 1995	
(see Limnocharitaceae	167)		Droseraceae	055		
Byttneriaceae			Ebenaceae	075		
(see Sterculiaceae	050)		Elaeocarpaceae	048		
Cabombaceae	013		Elatinaceae	046		
Cactaceae	031	18. 1997	Eremolepidaceae	105a		
Caesalpiniaceae	088	p.p. 7. 1989	Ericaceae	073		

Eriocaulaceae	184		
Erythroxylaceae	118		
Euphorbiaceae	115		
Euphroniaceae	123a		
Fabaceae	089		
Flacourtiaceae	056		
(excl. Lacistemaceae	057)		
(excl. Peridiscaceae	058)		
Gentianaceae	139		
Gesneriaceae	155		
Gnetaceae	209	9. 1991	
Gramineae			
(= Poaceae	187)	8. 1990	
Gunneraceae	093		
Guttiferae			
(= Clusiaceae	047)		
Haemodoraceae	198	15. 1994	
Haloragaceae	092		
Heliconiaceae	191	1. 1985	
Henriquesiaceae			
(see Rubiaceae	163)		
Hernandiaceae	007		
Hippocrateaceae	110	16. 1994	
Humiriaceae	119		
Hydrocharitaceae	169		
Hydrophyllaceae	146		
Icacinaceae	112	16. 1994	
Hypericaceae			
(see Clusiaceae	047)		
Iridaceae	200		
Ixonanthaceae	120		
Juglandaceae	024		
Juncaginaceae	170		
Krameriaceae	126		
Labiatae			
(= Lamiaceae	149)		
Lacistemaceae	057		
Lamiaceae	149		
Lauraceae	006		
Lecythidaceae	053	12. 1993	
Leguminosae			
(= Mimosaceae	087)		
+ Caesalpiniaceae	088)	p.p. 7. 1989	
+ Fabaceae	089)		
Lemnaceae	179		
Lentibulariaceae	160		
Lepidobotryaceae	134a		
Liliaceae	199		
(incl. Amaryllidaceae)			
(excl. Agavaceae	202)		
(excl. Smilacaceae	204)		
Limnocharitaceae	167		
(incl. Butomaceae)			
Linaceae	121		
Lissocarpaceae	077		
Loasaceae	066		

Lobeliaceae			
(see Campanulaceae	162)		
Loganiaceae	138		
Loranthaceae	105		
(excl. Viscaceae	106)		
Lythraceae	094		
Malpighiaceae	122		
Malvaceae	052		
Marantaceae	196		
Marcgraviaceae	044		
Martyniaceae			
(see Pedaliaceae	157)		
Mayacaceae	183		
Melastomataceae	099	13. 1993	
Meliaceae	131		
Mendonciaceae	159		
Menispermaceae	017		
Menyanthaceae	145		
Mimosaceae	087		
Molluginaceae	036		
Monimiaceae	005		
Moraceae	021	11. 1992	
Moringaceae	069		
Musaceae	192	1. 1985	
(excl. Strelitziaceae	190)		
(excl. Heliconiaceae	191)		
Myoporaceae	154		
Myricaceae	025		
Myristicaceae	003		
Myrsinaceae	080		
Myrtaceae	096		
Najadaceae	173		
Nelumbonaceae	011		
Nyctaginaceae	029		
Nymphaeaceae	012		
(excl. Nelumbonaceae	010)		
(excl. Cabombaceae	013)		
Ochnaceae	041		
Olacaceae	102	14. 1993	
Oleaceae	152		
Onagraceae	098	10. 1991	
Opiliaceae	103	14. 1993	
Orchidaceae	207		
Oxalidaceae	134		
Palmae			
(= Arecaceae	175)		
Pandanaceae	177		
Papaveraceae	019		
Papilionaceae			
(= Fabaceae	089)		
Passifloraceae	062		
Pedaliaceae	157		
(incl. Martyniaceae)			
Peridiscaceae	058		
Phytolaccaceae	027		
Pinaceae	210	9. 1991	

Piperaceae	009		Styracaceae	076	
Plantaginaceae	151		Suraniaceae	086a	
Plumbaginaceae	039		Symplocaceae	078	
Poaceae	187	8. 1990	Taccaceae	203	
Podocarpaceae	211	9. 1991	Tepuianthaceae	114	
Podostemaceae	091		Theaceae	043	
Polygalaceae	125		(incl. Bonnetiaceae)		
Polygonaceae	038		Theophrastaceae	079	
Pontederiaceae	197	15. 1994	Thunbergiaceae		
Portulacaceae	034		(see Acanthaceae	156)	
Potamogetonaceae	171		Thurniaceae	185	
Proteaceae	090		Thymeleaceae	095	
Punicaceae	097		Tiliaceae	049	17. 1995
Quiinaceae	045		Trigoniaceae	124	
Rafflesiaceae	108		Triuridaceae	174	5. 1989
Ranunculaceae	015		Tropaeolaceae	135	
Rapateaceae	181		Turneraceae	061	
Rhabdodendraceae	086		Typhaceae	188	
Rhamnaceae	116		Ulmaceae	020	11. 1992
Rhizophoraceae	101		Umbelliferae		
Rosaceae	084		(= Apiaceae	137)	
Rubiaceae	163		Urticaceae	023	11. 1992
(incl. Henriquesiaceae)			Valerianaceae	165	
Ruppiaceae	172		Velloziaceae	201	
Rutaceae	132		Verbenaceae	148	4. 1988
Sabiaceae	018		(incl. Avicenniaceae)		
Santalaceae	104		Violaceae	060	
Sapindaceae	127		Viscaceae	106	
Sapotaceae	074		Vitaceae	117	
Sarraceniaceae	054		Vochysiaceae	123	
Scrophulariaceae	153		Winteraceae	001	
Simaroubaceae	130		Xyridaceae	182	15. 1994
Smilacaceae	204		(incl. Albolbodaceae)		
Solanaceae	142		Zamiaceae	208a	9. 1991
Sphenocleaceae	161		Zingiberaceae	193	1. 1985
Sterculiaceae	050		(excl. Costaceae	194)	
(incl. Byttneriaceae)			Zygophyllaceae	133	
Strelitziaceae	190	1. 1985			